AURORA

A celebration of the Northern Lights

Greatland Graphics
Anchorage, Alaska
www.alaskacalendars.com

AURORA
ISBN-10 0-936425-962
ISBN-13 9780936425962
Published and distributed by:
Greatland Graphics
PO Box 100333
Anchorage, Alaska 99510
www.alaskacalendars.com
Printed in China

Cover photo: Aurora over Knik
River, Alaska, by Cary Anderson.

Title page: Rare pink and multi-
colored aurora expands across
the stars at zenith directly
overhead, filling the opening in a
spruce and birch forest near
Talkeetna and overpowering the
moon, by Dave Parkhurst.

Dedication page: Aurora
illustration by Colin Lemar,
© Dave Parkhurst

Dedicated to Martin Brendel
(1862-1939)
The first photographer to capture the aurora in 1892
and
to our wives, for their winter nights alone

Nocturnal Encounters

Between us, Dave Parkhurst and I have spent more than 40 winters stalking Alaska's icy darkness in pursuit of the aurora borealis. In the valleys and mountains of our nocturnal haunts, silent nights have been punctuated by owls hooting, moose grunting and strange animal noises we could not identify.

On a handful of occasions, the sudden commencement of a bright and shifting aurora has caused a distant pack of wolves to bay. This was no strange coincidence. We have heard wolves howl at least a half dozen times when the aurora burst forth with ample brilliance to distract or excite any terrestrial being.

Bald eagles, too, awakened and shrieked from their

(facing page) **The aurora appears during a midnight drive on the Alaska-Canada Highway near Beaver Creek, Yukon Territory.**

(right) **A wolf howls at the aurora in this painting, circa 1915, by famed Alaska artist Sydney Laurence.**

5

nighttime roosts in old cottonwoods when the northern lights rippled intensely across the sky. We were surprised and delighted to discover that wild creatures acknowledge the aurora.

Dave and I have logged hundreds of bone-chilling nights, shuffling along on crusty snow and working by flashlight. We always monitor the periphery of our photo locations, occasionally sweeping with our flashlight beams to check for the reflective glow of peering animal eyes. This is especially important during spring and fall when Alaska's bears are awake and moving about.

Although we have often seen bear tracks along riverbanks, glowing eyes have almost always turned out to be a moose, fox, beaver, or porcupine. *Almost* always.

One evening after sunset near Talkeetna, Dave and I set up our camera gear in a small clearing along a wooded bluff with a panoramic view of Mt. McKinley and the Alaska Range. I leaned on an old spruce stump

(left) **Aurora dances above Mount McKinley as seen from the south side.**

(facing page) **Aurora behind the "world's largest gold pan" at the Kluane Museum of Natural History, Burwash Landing, Yukon Territory, Canada.**

6

(above) **A red fox wanders past a cozy cabin beneath the northern lights in this stamp from Finland. According to Finnish legend, the heatless fire of the aurora borealis is started when the fox strikes his tail on the snow.**

(right) **Cary Anderson photographs and watches an auroral glow, 50 miles north of Anchorage.**

while Dave sat on a fallen log about 15 feet away.

We quietly hoped for some aurora action, while hours passed without exposing any film. The scenery was delicately lit by the moon, the autumn sky was crystal clear and the temperature was a comfortable 25 degrees. Conditions for night photography were ideal, but the northern lights were not cooperating.

"Crack." A small branch snapped in the distance behind us. Dave and I turned to look at each other with raised eyebrows in the dim moonlight. "Snap." The sound of rustling leaves and breaking twigs in the brushy undergrowth was slowly coming nearer.

Although neither of us could see the source of the noise, Dave whispered urgently,

"It's a bear!"

"Oh, come on," I answered skeptically, "it's just a moose."

"No." Dave insisted, "*That* sounds like a bear."

I kidded Dave about his alleged ability to distinguish species by the sounds of sticks breaking. "Relax," I chuckled, "not everything crunching through the woods at night is a bear. There's no way you can tell the difference. No way."

The crunching and snapping noises became louder

as they drew closer. Something was coming. Now, we were standing straight up. Dave clicked on his flashlight, pointing it in the direction of the mystery beast.

No eyes. Nothing.

For a few minutes, the noises stopped. I, too, drew a flashlight from my pocket and focused its beam. Neither of us spoke. We squinted intently at the edge of the brush about 30 feet away.

Then, as if responding to a backstage cue, the hulking form of a bear emerged in the center of our spotlight. It was a grizzly. Its glowing eyes were fixed on us. Dave and I had seen plenty of bears in our Alaska travels, but always at a fairly safe distance in broad daylight. This was the heart-stopping moment in the darkness we always thought might come.

Neither of us carried a gun or pepper spray, though many nights we talked about doing so—just in case. Now we were face to face with a grizzly. Either by intent or coincidence, it had found us. Did the bear

(left) **A band of aurora illuminates the sky over a campsite in the Chugach Mountains.**

(right) **A weather-contorted spruce, just east of Alaska's Gunsight Mountain, is silhouetted by a delicate aurora.**

(top) **A stylized aurora appears in the background of a 1954 "large letter" Alaska postcard.**
(bottom) **Viewed from Alaska's Montague Island, an aurora swirls over Patton Bay in Prince William Sound.**

track us with the intent to eat us?

Our adrenalin was mixed with a twinge of dread. The bear was too close. *Way* too close.

Both of us began shouting, "Hey! Hey, bear!" as we clanked our flashlights on our metal tripods expecting the bear to flee instantly. It did not. It just stood motionless, with an adversarial posture, staring into our lights.

We kept our lights pointed at the bear's eyes as much as possible, hoping to temporarily impair its night vision. A half-blind bear might have a harder time mauling us. After what seemed like 2 minutes, but was probably only 8 seconds, the grizzly turned sideways and casually waddled away.

Unfazed by our yelling and clanking, the bear paused once to glance at us over its shoulder before disappearing back into the darkness. When the bear was safely distant, Dave and I sputtered exclamations that are better left unprinted.

Our nocturnal encounter with the grizzly ended peacefully, and Dave established himself as an expert on things that go "crunch" in the night. After sunrise, we returned to our homes with a hair-raising tale for friends and family, but not a single aurora photo.

The aurora takes on a serpentine shape just after sunset near Savage River in Denali National Park.

Of Legend & Lore

Bearing forth theory or ideology, humans have never been without explanation for the mysteries of our planet. As the earth was once believed to be flat and the oceans menaced by sea monsters, the aurora, too, has been subject to fanciful interpretations.

All northern cultures acknowledged the aurora—some with veneration, others with dread. According to one well-known Eskimo superstition, a person should not dare to whistle in the presence of an aurora. To do so might cause the aurora to descend and slice-off the whistler's head.

Eighty-three-year-old Ada Ward, a native elder born and raised in Kotzebue, Alaska, says all the kids

(facing page) **Aurora and moonrise play over the northern Chugach Range making the sub-zero delta sparkle. Ice fog rises from open leads in the river.**

(right) **Aurora on the horizon of the Bering Sea lights the background of a walrus skin boat on a scaffold at Gambell, St. Lawrence Island, Alaska.**

16

AURORA BOREALIS.

From a Sketch in Lapland by KARL BOCK.

in her village along the Chukchi Sea were told, "if you whistled and hollered outdoors, the northern lights would come down, take your head off and use it for a football." During her childhood, Ward scrambled home as fast as she could when the aurora was active.

In the village of Barrow along the Alaska's Arctic coast, tossing frozen dog feces into the air was once thought to be effective at keeping the aurora's mean spirit away. Some believed carrying a knife offered some protection. Others thought they could force the aurora to retreat by clapping their hands.

Not all indigenous northerners feared the aurora. Some believed the aurora was akin to heaven. After death, persons who were generous and good-hearted ascended to the aurora where the light is always bright and the hunting is easy. Others called on the aurora for medicinal purposes.

Eskimos from Siberia to Greenland embraced the notion that auroras were the spirits of dead ancestors who were playing some sort of celestial ball game with a walrus skull. On Alaska's Nunivak Island in the Bering Sea, villagers looking skyward had an opposite interpretation. They saw walruses playing ball with a human skull.

(above) **An Eskimo boy and his puppy sit beneath the northern lights on the cover of a 1936 Alaska postcard folder.**

(facing page) **A rayed arch shines above some peculiarly skinny walruses in this antique lithograph, circa 1870.**

Northern lights dance on the cover of a 1945 booklet about dairy farming in Alaska. Illustrated by Dwight Mutchler.

Journals from anthropologists, explorers, missionaries and traders suggest that the Eskimos and other Native Americans offered countless explanations and beliefs about the aurora; so many, in fact, that they defy generalization. Each tribe, clan or village gave a slightly or vastly different tale or account.

Author Richard Nelson, a cultural anthropologist, reports that Athapaskan natives living along the Yukon and Koyukuk rivers believed the aurora came from the spirit of a man who broke his bow while hunting and, later, died in a fire.

In his book, *Make Prayers to the Raven*, Nelson recounts, "When the aurora runs in brilliant curtains across the night sky, trembling and flashing with a glow that illuminates the landscape, it is Northern Lights Man shooting his arrows into the heavens."

Among the more fantastic legends is one attributed to the Makah Indians of Washington state. They believed the aurora came from fires built by a tribe of dwarfs in the far north. According to a chapter written by Dorothy Jean Ray in *The Alaska Book*, these dwarfs, whose height was only half the length of a canoe paddle—but strong enough to capture whales with their bare hands—used the fires for boiling blubber.

19

(above) (A) The Canadian Air Force uses a green aurora over a red maple leaf on its Aurora Tactical Instructor patch. (B) In 1979, Boy Scouts from the Tomahawk District in Texas sported a polar bear and aurora patch on their uniforms. (C) A Boy Scout patch celebrates a 1994 pow wow between the Georgia and Alaska councils.

Another legend credits exceptionally *big* people for the mysterious lights. "The Menomee Indians of Wisconsin," Ray wrote, "regarded the lights as torches used by great giants in the North, to spear fish at night."

Some central Canadian Indians had at least some scientific basis for their explanation of the aurora. They believed the aurora was caused by herds of deer in the sky. This notion is thought to have been inspired by the static discharge that occurs when stroking deer hair.

Eskimos and Indians weren't the only ones whose imaginations helped to define the aurora. Europeans also had ideas about the northern lights that seem quite strange compared to what we know today. Swedes once referred to the aurora as *sillblixt*, which translates as "herring flash," because they thought giant schools of the fish turning in unison were causing the sky to glow.

(left) **Fingers of brilliant green aurora reach skyward above a frozen channel just east of Knik Arm in southcentral Alaska.**

(facing page) **Multi-colored aurora dances above the Knik River.**

Misunderstood Aurora

Historically, people have often been startled by their initial encounter with the northern lights. When the aurora dipped south and made rare appearances in Europe's middle latitudes, citizens were stricken by the fear of God. In his 1897 book, *The Aurora Borealis*, French meteorologist Alfred Angot reported widespread panic:

> The aurora borealis had become a source of terror.... At the sight of them people fainted... others went mad. Pilgrimages were organized to avert the wrath of Heaven, manifested by these terrible signs. Thus, according to the Journal of Henry III, in the month of September 1583, eight or nine hundred persons of all ages and both sexes, with their lords, came to

(left) **A snowbound cottonwood tree is silhouetted by a ruddy aurora in Arctic Valley, Fort Richardson, Alaska.**

(right) **Purplish hues of aurora glow behind a snow-covered spruce in the mountains just beyond Anchorage.**

(above) Santa listens to the radio with the northern lights above his igloo in a promotional postcard, circa 1909, for the North Pole Wireless Company.

(facing page) A fiery, red and yellow aurora reflects in a small, mountain stream.

Paris in procession, dressed like penitents or pilgrims…to say their prayers and make their offerings in the great church at Paris; and they said they were moved to this penitential journey because of signs seen in heaven and fires in the air….

The appearance of the northern lights continued to trigger alarm for centuries, such as in the winter of 1938 when a brilliant display was seen over England. The London bureau of the Associated Press reported, "The ruddy glow led many to think half the city was ablaze." Firemen rushed to Windsor Castle because they thought it was burning. The story which ran in the *New York Times* also noted that the aurora "spread fear in parts of Portugal and Austria…."

Another end-of-the-world scenario was reported by author Neil Davis in *The Aurora Watcher's Handbook*. The author's footnotes offer a first-hand account of a red aurora that covered the sky above a small, West Virginia coal mining town in 1941:

Most of the town's residents believed the end was nigh so they ran into the streets where they entered into group prayer, religious songs and a recanting of past sins. The most memorable part of it all was the confession of a young neighbor lady who apologized at the top of her voice for what to my youthful ears were some highly interesting social activities.

26

An increased awareness of the northern lights has probably reduced the perception of doom often associated with them, due to the fast dissemination of news and information in the 21st century. But even today, the aurora apparently is feared by some.

Auroraphobia is a clinical term used to describe people inflicted with a persistent, irrational fear of the northern lights. Evidently, the affliction is quite rare. Numerous calls to psychology clinics in Alaska and elsewhere, during preparation of this manuscript, failed to turn up a single therapist who has treated a case of auroraphobia.

(facing page) **The center of night bursts with color, expanding like a bird's wings, as a rare and powerful auroral corona fills the winter skies above Glennallen.**

(right) **A French trade card promoting Liebig's canned corned beef in 1925 depicted the aurora over an Arctic adventure scene.**

28

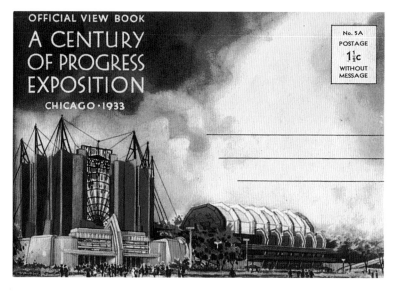

(above) At Chicago's 1933 Century of Progress Fair, an attempt was made to simulate the aurora borealis. According to the *Chicago Sun-Times*, the northern lights were simulated by "shining lights on fog created by chemicals dropped from airplanes."

(left) The sun rises over Alaska's Cook Inlet.

(facing page) Reds, greens and yellows appear in a multi-hued auroral display above the Knik River.

The Sun's Gift

Ninety-three million miles away, we find the power generator for the aurora. The same sun which lights our day and provides for life on earth also causes the nocturnal glow of the northern lights.

Gigantic, fiery flares on the surface of the sun send streams of electrically-charged particles hurtling toward earth at speeds sometimes exceeding a million miles per hour. These soaring electrons, known collectively as the solar wind, collide with our planet's magnetic field or "magnetosphere."

Dr. Carl Gartlein wrote in a 1947 *National Geographic* article that "as the sun revolves around its axis, these huge streams sweep around through space much as do the jets of water issuing from a rotating lawn sprinkler. Every so often one of them catches the earth in its path…." These collisions with earth's magnetic field trigger the production of light-emitting photons which appear as the glow of the aurora.

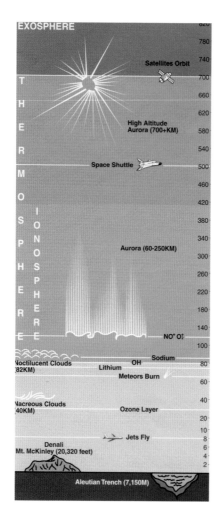

The northern lights are created much the same way a neon sign is lit. A stream of electricity directed at the gaseous contents within a neon tube causes light. Color variations depend on the gases within. A similar principle creates the images we see on our television screens.

The color of the aurora depends entirely on which gases in the upper atmosphere are struck by the solar wind. As energized particles rain down on the earth's magnetic field, oxygen atoms emit the color green. Higher altitude oxygen creates rare, red auroras. Ionized nitrogen molecules are responsible for blues, while neutral nitrogen gives us purples.

Green is the most common color of aurora. But anyone who is vigilant enough, or lucky, may witness an exciting range of color. Countless journalists and writers have testified as to the dazzling intensity and variety of the northern lights.

Sara Machetanz described a display which "lit up the sky like the 4th of July." In her book *Where Else But Alaska?*, she marveled at how "streamers of pink,

Auroras are found in the ionosphere in a band roughly 70 to 200 miles high. They have been photographed from above by satellites and space shuttles.

During the polar winter when night is 20- to 24 hours long, the aurora may make a complete circle (known as the auroral oval) around the polar hemisphere.

While auroras have terrified or delighted observers for thousands of years, space exploration and other advances in technology have allowed scientists to better explain them.

In Roman mythology, "Aurora" is the Goddess of Dawn as depicted in a steel engraving, circa 1865.

lavender, green and yellow arched to the zenith of the heavens."

In *Ballad of the Northern Lights*, famed gold rush era poet Robert Service wrote, "And the skies of night were alive with light, with a throbbing thrilling flame; Amber and rose and violet, opal and gold it came."

Auroras occur just beyond the earth's atmosphere at the edge of space. The altitude of the aurora has been measured as low as 40 miles above the earth to as high as 400 to 600 miles during periods of extreme auroral activity. Typically, however, auroras occur 60-to 120 miles above us.

Viewing the aurora is strictly a night activity, but—truth is—the phenomenon occurs 24 hours a day. Like the stars, the aurora—no matter how bright or persistent during the night—will be drowned by the overpowering light of dawn.

According to Alaska Geographic's *Secrets of the Aurora Borealis*, by Dr. Syun-Ichi Akasofu, the aurora gets it name from "Aurora, the rosy-fingered goddess of the dawn in Roman mythology who heralded the rising sun." Aurora scholars favor Italian physicist Galileo as being responsible for the name "aurora borealis."

(left) **A huge dome of aurora caps the sky over the Copper River Basin and the mighty Copper River, as sub-zero temperatures begin to freeze the waters and grip the region in an Alaska winter.**

(above) **The aurora australis shines energetically behind Admiral Richard Byrd in a 1958 advertisement for John Hancock Insurance.**

The list of contributors to theories about the northern lights reads like a who's who of the scientific world. Aristotle, who wrote one the first scientific accounts of the aurora, thought vapors rising from the earth caught fire in the upper atmosphere. Sir Edmund Halley—best known for the comet bearing his name—surmised a connection with magnetism, while Benjamin Franklin correctly speculated on the aurora's electrical component.

Even as auroral science matured, a handful of theorists still entertained wild ideas about the aurora's origin. In his 1906 book, *Phantom of the Poles*, William Reed argued that the earth was hollow and the aurora vented through holes at the north and south poles. Some subscribed to this theory despite the growing scientific evidence against it.

Near the turn of the 20th century, scientists were piecing the aurora puzzle together. Norwegian professor Kristian Birkeland hypothesized that the northern lights were created when charged particles from the sun were caught by the earth's magnetic field. By shooting electrons at a magnetic ball inside a glass vacuum chamber, he was able to imitate the aurora in his lab.

35

(above) **The northern lights are depicted on a gold pan with a musher and dog team in the Yukon.** (Painting by E. Nowlan.)

(facing page) **Dusk silhouettes the interior peaks of Denali National Park as a rising moon lights a rock face and the East Fork Toklat River below.**

UNDER NORTHERN LIGHTS

Far-fetched? Not a bit of it, except in the sense that this box of Jell-O has been brought a long, long way. For we do have customers who live under the Arctic Circle, and who say cold, hard things of us if we do not arrange for shipping connections before the trails are closed with the winter's snows.

RASPBERRY JELL-O

(above) **A 1922 recipe booklet proclaims that Jello is "At Home Every Where," even in the arctic under the northern lights.** (facing page) **Comet Hale-Bopp and the northern lights appeared together over the Knik River Valley in March 1997.**

Teams of scholars—especially those in northern nations—continue to study the aurora at observatories and remote outposts. They have even succeeded in creating artificial auroras. Combining rocket science and pyrotechnics, scientists at the Poker Flat Research Range in Alaska have simulated auroras by launching rockets that expel barium vapor at the edge of space.

Established by the Geophysical Institute of the University of Alaska—Fairbanks, the Poker Flat facility is uniquely dedicated to studies of the aurora borealis and other atmospheric phenomenon. In recent years, Institute scientists have published routine aurora forecasts in local newspapers and on the internet.

While much is now known about the aurora's behavior, the mystery endures as to whether it makes an audible noise. Witnesses have often claimed that the northern lights make a sound. Hundreds of testimonials on auroral noise, which describe whooshing, whistling, crackling or humming, have been filed in scientific archives.

Some have likened the sound to a static discharge, or the sizzle from a frying pan. A weather observer in Alaska once described the sound as being similar to a

spitting cat. In other accounts, the aurora made a hissing sound which caused dogs to growl and bark.

Nearly all noise reports are associated with auroras that are bright and energetic. Most ear-witnesses have also claimed the sounds were synchronized with the rapidly surging or sweeping motion of the lights. Despite reports from many credible sources, the subject of auroral sound remains a matter of scientific debate. To date, nobody has been able to record it.

One explanation offered by scientists is that the noise may come from an electrostatic reaction in the composition of certain rocks and soil. Another possibility is that dental fillings in a person's mouth may act as a receiver for electromagnetic waves during high aurora activity.

The latter could account for the perception of sound, as some people have been known to pick up radio stations through the metal in their mouths. The aurora is, after all, according to Fairbanks physics professor Dan Swift, "the world's most powerful radio station."

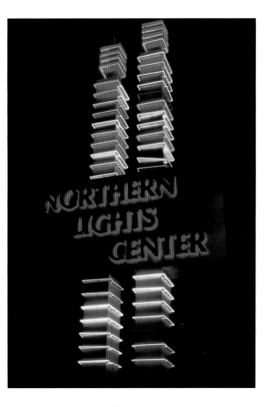

(above) **A neon sign illuminates one of Anchorage's major shopping centers.**

(facing page) **Red aurora dominates the sky in a wide-angle view of Pioneer Peak from the Matanuska Valley.**

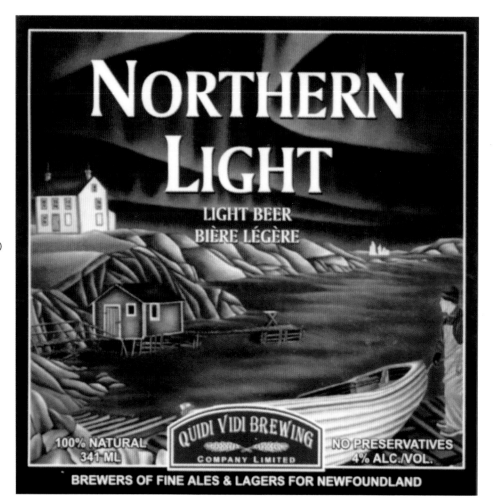

(above) **A polar bear enjoys a cold beverage in this stylish logo for Aurora Vending of Alaska.**

(left) **Aurora adorns the sky above a Newfoundland seaside village on a contemporary beer label.** (Art by Danielle Loranger for the Quidi Vidi Brewing Company.)

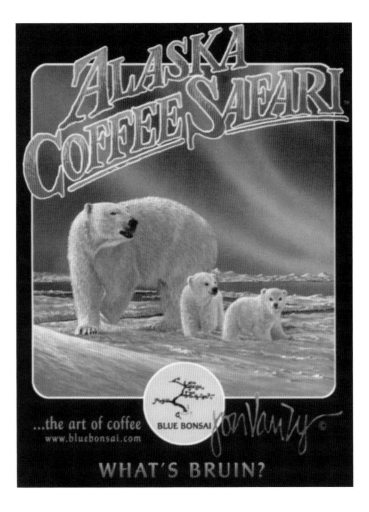

(above) **A 1950s Norwegian Advertising brochure for furs.**

(left) **An Arctic night with the aurora borealis and a polar bear with cubs adorns a coffee label.** (Art by Jon Van Zyle.)

Powerful Energy

The "electroject" that induces the northern lights is said to be 1 million megawatts strong and has been known to charge the 800-mile Trans-Alaska Oil Pipeline with 100 amps of electricity.

Such power has long been apparent, especially in the communications business. In 1852, a superintendent at the Boston and Vermont Telegraph Company reported that his instruments continued to make lines on telegraph paper — 3 hours after the batteries were removed. The phantom power was attributed to "the strong and brilliant Aurora of that day."

Solar flares, especially those that produce powerful auroras, continue to affect telecommunication. Satellites, radios, TVs and telephones are prone to malfunction when solar storms erupt. Power outages also may occur, the most famous of which was on March 13, 1989 when an electrical blackout darkened the city of Quebec.

(above) **Powerful solar storms trigger beautiful auroral displays, but can cause problems with telecommunications and electric power transmission.**

(facing page) **One of the largest auroral displays on record glows powerfully over Pioneer Peak and Twin Peaks south of Palmer with a rare pink aurora. This event plunged Quebec, Canada into total darkness after causing a massive power grid failure.**

43

44

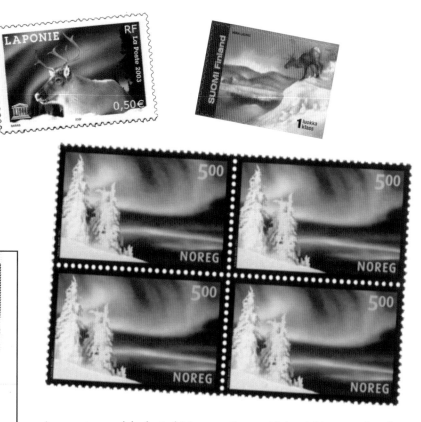

LAPONIE   RF   La Poste 2003   0,50 €

SUOMI Finland   1 luokka klass

EUROPA 91 · RYMDEN

SVERIGE 4 kr   SVERIGE 4 kr   SVERIGE 4 kr

EUROPA 91 · RYMDEN
Förstadagsbrev · First Day Cover
1991:4

Herr Erik Herz
Jens Kofodsgade 4
DK-1268 Köpenhamn

Aurora stamps: (clockwise) Norway, France, Finland, Norway, Sweden.

(facing page) Ice fog creates a lingering lunar halo as a ruby red aurora flames up to the east beyond Prince William Sound. Fox tracks lead off into the frosty brush.

North & South

The same forces that deliver the aurora borealis in the northern hemisphere also produce the aurora australis, known as the southern lights, in the southern hemisphere. Aurora polaris, or "polar lights," are terms sometimes used to describe either or both, although the shorter "aurora" is most common.

Somewhere on earth, the aurora is visible every night, most commonly occurring in doughnut-shaped ovals above the far northern and southern hemispheres. Fluctuations in solar activity influence the size of the auroral ovals. Intense solar winds and geomagnetic forces cause expansion and shifting of these ovals and, thus, may cause the aurora to be visible over a greater area.

The northern and southern lights happen at the same instant and are nearly identical. Scientists using

(above) **Icebergs, aurora and a polar bear are depicted on a 1900s Belgian chocolate label.**

(facing page) **The aurora australis, or southern lights, are seen at the bottom of the world in this view captured during the Challenger/Spacelab III mission.**

A 1928 Will's Cigarette card portrays penguins enjoying a view of the aurora australis. Educational and collectible, the cards were given away as premiums.

48

special, strategically-placed cameras have determined that the auroral light patterns in the northern hemisphere coincide with those in the southern hemisphere. A mirror-like image of the aurora may be seen when comparing photos taken, simultaneously, in opposite hemispheres.

Auroras in the southern hemisphere are witnessed by fewer people due to sparse populations living within the auroral zone. Nonetheless, the southern lights have generated plenty of scientific fascination and personal appreciation. Polar explorer Robert F. Scott, who died from cold and starvation in the Antarctic wilderness, wrote this entry in his journal in 1912:

It is impossible to witness such a beautiful phenomenon without a sense of awe, and yet this sentiment is not inspired by its brilliancy but rather by its delicacy in light and colour, and above all by its tremulous evanescence of form. There is no glittering splendor to dazzle the eye, as has been too often described; rather the appeal is to the imagination by the suggestion of something wholly spiritual, something instinct with a fluttering ethereal life, supremely confident yet restlessly mobile.

(above) A postcard mailed from Moscow depicts an auroral curtain above the Soviet nuclear icebreaker, "Lenin." (Art by S. Pomanski.)

(right) A single auroral curtain sweeps southward across the stars of Orion, with moonset highlighting the snow-covered peaks in the heart of Denali National Park.

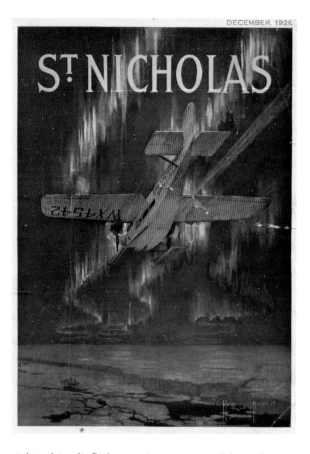

Admiral Byrd's flight over Antarctica is celebrated on the cover of St. Nicholas magazine in 1928.
(Illustration by Howard Brown.)

In the 1930's, American explorer Admiral Richard Byrd spent a winter alone in Antarctica where he, too, wrote a stirring account of his experience watching the aurora australis:

> Overhead, the Aurora began to change its shape and become a great, lustrous serpent moving slowly across the zenith. The small patch in the eastern sky now expanded and grew brighter; and almost at the same instant folds in the curtain over Pole began to undulate, as if stirred by a celestial presence. Star after star disappeared as the serpentine folds covered them. .... I was left with the tingling feeling that I had witnessed a sight denied to all other mortal men.

Ninety-nine-year-old Colonel Norman Vaughan, who was the chief dog driver for Byrd's first Antarctic expedition in 1928, recalls being inside a building at the base camp when someone excitedly announced, "the northern lights are out!" to which he replied, "no, those are the 'southern' lights." He described the aurora australis as being "bright and colorful," but no different from the aurora borealis.

Vaughan had previously seen the northern lights, so he spent little time observing the southern version. In the sub-zero Antarctic cold, he says, "you looked

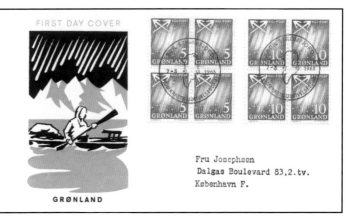

(above) **Books on polar exploration often include vivid passages describing the aurora borealis or aurora australis.**

(above, right) **Patch worn by the Alaska Royal Rangers, a ministry of the Assembly of God. Quyurrtequ is a Yup'ik Eskimo word meaning "the gathering."** (Design by Harlan Legare.)

(right) **Stamps on postal cachet: Greenland**

Postage stamps: Tanzania (1) Australia (2)

52

quickly, and after you saw them, you put your hat back on. We said, 'gee, they're beautiful' and went back inside."

Russell Owen, a reporter for the *New York Times* who visited with Admiral Byrd in Antarctica, described the southern lights in his 1952 book, *The Conquest of the North and South Poles.* He recalled, "Sometimes it resembled curtains that rippled in the sky; at other times ropes that twisted upon themselves. Once I saw an aurora that was almost overhead, spinning around like a giant pinwheel."

Polar explorers and adventurers at both ends of the earth eagerly shared their visions of the aurora, often with poetic flare. In 1911, Frederick Cook described the aurora he saw on an expedition to the North Pole:

> We continued our course, the Eskimos singing, the dogs occasionally barking. Hours passed. Then we all suddenly became silent. The last, the supreme, glory of the North flamed over earth and frozen sea. The divine fingers of the aurora, that unseen and intangible thing of flame, who comes from her mysterious throne to smile upon a benighted world, began to touch the sky with glittering, quivering lines of glowing silver.

Cook expressed how insignificant he felt in the presence of the northern lights and concluded, "Spiritually intoxicated, I rode onward. The aurora faded. But its glow remained in my soul."

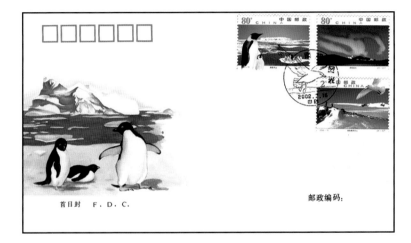

(above) Aurora australis appears on a stamp in the upper right corner of a Chinese First Day Cover depicting Antartica.

(right) Wild auroral curtains undulate over the Chugach Mountains with Ursa Major, the Big Dipper. Open water from spring melt reflects the display.

54

Flirt With Phenomena

The awe-inspiring aurora is drawing increased appreciation by adventure travelers eager for a new, off-beat experience. Some locales in Alaska and Canada favored by frequent auroral activity are experiencing a boom of aurora tourists. Despite temperatures which may plunge to -50 degrees Fahrenheit in mid-winter, Alaska and northern Canada have become a hot attraction.

Less than a two decades ago, the far north was almost exclusively a summer destination. Vacation lodges were boarded-up and nailed shut during the long, cold winter due to a lack of visitors. Twenty or

(facing page) **Red and yellow aurora glows beside the moonset, silhouetting the 20,320- foot Mount McKinley as viewed from Denali State Park.**

(right) **The northern lights appear over a home in Alaska's Matanuska Valley.**

more hours of darkness cloak the state's northern half during the shortest, most frigid days of the year. Closing down in winter was taken for granted.

In an article for the *New York Times* in 1992, Timothy Egan noted the difficulty of luring visitors to Alaska during winter. "Going outside requires enough insulation to pass the Anchorage building code. The bears are all hibernating, the moose are hungry and prone to cranky behavior, and the salmon are months away from the first big river run." But, he acknowledged, "Alaskans have figured out a way to sell the sky."

Many lodge proprietors are now open year-round, scrambling to keep pace with the off-season demand for rooms and aurora tours. Telephone and mail inquiries to the Alaska Division of Tourism about the northern lights come from all across the United States and a number of foreign countries. Most aurora seekers, however, come from Korea and Japan.

It has often been reported that Japanese are drawn to the aurora because they believe conceiving a child

Aurora graces the heavens behind a Russian Orthodox Church in the Indian village of Eklutna, Alaska.

under the northern lights will bring good fortune. "That's a myth," says Deb Hickok of the Fairbanks Convention and Visitors Bureau. "It's a fabrication, and many Japanese find it highly insulting." They come to see the aurora, she says, because of their affinity for the beauty of nature.

Regardless of motives for aurora viewing, Fairbanks, Alaska is among the preferred destinations for aurora seekers. The northern lights may be viewed across much of the state, but Fairbanks lies near the auroral oval where the odds of seeing them are good on most winter nights. The city's location in Alaska's interior also offers clear skies more often than in coastal communities.

Coldfoot, north of Fairbanks at mile 175 of the Dalton Highway, lies directly beneath the auroral oval and offers a more remote experience. Kathy Hedges, reservations coordinator at Coldfoot Camp, says the lodge offers some advantages over the city. "You can just go out the back door and be away from artificial lights."

Others in the Fairbanks area catering to aurora tourists include Chena Hot Springs Resort, Northern Sky Lodge, and of course, the Aurora Borealis Lodge.

Poster promoting Fairbanks features a paddlewheel steamboat and the aurora above Mount McKinley.

57

During the day, hearty visitors will find plenty to do. Local tour operators offer cross-country skiing and dog mushing.

In winter, Dave Parkhurst leads Aurora Watch® Tours to Talkeetna and other Alaska locations for those wanting a personal aurora experience with a knowledgeable guide.

Canada, Greenland, Iceland, Scandinavia and even northern Scotland are sharing in the trend toward winter tourism, thanks in part to the northern lights. "Aurora tourism" is now a marketable and recognized phrase in the northern visitor industry.

(left) **Puffy backlit clouds attempt to hide a coastal aurora as water, earth and sky came together for only a moment.**

(facing page) **Aurora rises over the eastern peaks of the Alaska Range as fog floats above the Tangle Lakes along the Denali Highway.**

(left) A frost-covered Nikon F5 camera illustrates the cold weather conditions on a typical night of aurora photography.

(right) Aurora photographer Cary Anderson, bundled up for extreme cold, contemplates the unfolding scene, anticipating a great night ahead.

(facing page) Subtle aurora dances above the crescent moon on an outhouse door at Tangle Lakes Lodge on Alaska's Denali Highway.

Shooting The Glow

Through many years of stumbling around in the dark, Dave and I have become excruciatingly familiar with the most basic principal of aurora photography: a stubbornness to persevere through toe-numbing cold is far more important than the technical aspect.

Recording the phenomenon on film is simple. If you own an SLR camera with a standard or wide angle lens, a shutter release cable and a tripod, you are technically ready to shoot an aurora. Anyone who has ever used the bulb setting on a manual camera to expose a frame of film for longer than a second is adequately trained. The same is true if you've used a computerized camera to control a multi-second exposure.

All of the photographs in this book were shot on film, however, digital cameras give photographers one distinct advantage: the ability to see and compare results from different camera settings. Viewing your

results in the field can allow you to make immediate adjustments for a more accurate exposure.

Exposure times for shooting an aurora will vary widely depending on a handful of unique factors. No two auroras are the same. The color and intensity of an aurora can vary from minute to minute. It may range from a faint, barely discernable glow to a dazzling storm of eye-popping color. In fact, the northern lights will often brighten or dim during your exposure. A light meter is worthless under these circumstances.

Consider, too, that an aurora is typically an evolving shape. Some auroras slowly expand and contract without migration. Others writhe across the sky like a snake. In fact, the northern lights can slither right out of your frame before you finish your exposure. Shooting the northern lights is "simple," but not necessarily

(left) **Aurora and moonset silhouette Sleeping Lady (Mount Susitna), casting an orange winter light on the icy Knik Arm near Anchorage, Alaska.**

(opposite page) **An aurora is reflected in the ice and water in the Twentymile Valley near Portage. The ghostly trees are reminders of the 1964 earthquake; saltwater killed the trees when the land subsided.**

64   Lights from a passing vehicle illuminate the sturdy spans of the Nenana River Bridge as a soft auroral arc appears on the horizon.

easy. Some general guidelines can be applied for best results.

An aurora exposure may range from 3 to 100 seconds or longer. Faster is better to maintain the integrity of the aurora as you saw it, but your equipment and film of choice will dictate your own results. Dave and I have experimented with a wide variety of film types ranging from Kodachrome 64 to Fujichrome 100/1000.

Since our first attempt at shooting auroras more than 20 years ago, our film choices have varied widely, and still do. We shoot Kodak and Fuji films, almost exclusively, but so many new emulsions have popped up during the past decade that we have yet to agree on a favorite. Our best pictures have been made with 100, 200 and 400 speed films or equivalent pushes.

Exposing 200 speed film at f2.8 from 20 to 30 seconds will often produce acceptable results, but feel free to hold the shutter open longer if the aurora appears dim. A ballpark exposure can be extrapolated for other film speeds and f-stop settings. Remember that bracketing widely, *very* widely, is the name of the game when coaxing the aurora to stick to your film.

Remember that extremely cold weather can drain

the power from your batteries. For battery-dependent cameras in cold weather, it is always a good idea to carry an extra set in a warm interior pocket. Once equipped, you'll need a dark, clear sky and plenty of patience.

Dave and I have spent countless nights gulping hot, caffeinated beverages while staring with bloodshot eyes into the dark, only to watch dawn break without taking a single shot. Many times we have danced on the frosty tundra, not to rejoice, but to pump some warm blood into our half-frozen extremities.

Even in Alaska, a colorful light show is not guaranteed. If you are prepared to wait, your patience will be richly rewarded.

Aurora pins:
(top) The aurora glistens above a Russian ship. Words on the government-issued pin translate in English as, "In commemoration of service in the North Fleet."

(center) Alaska's largest winter carnival, the Anchorage Fur Rendezvous, featured an aurora over a cabin on its 1995 collector's pin. (Design by Barbara Stevens.)

(bottom) An enamel pin celebrating the 1987 Iditarod Trail Sled Dog Race depicts a musher beneath the aurora borealis.

66

67

(facing page) An auroral storm swirls behind the ski slopes of Alaska's Arctic Valley near Anchorage.

(left) This view of Saturn taken by the orbiting Hubble Space Telescope proves that the aurora phenomenon also occurs on the north and south poles of other planets. Similar photos have been taken of Jupiter.

(above) A Matanuska Valley farm scene is decorated by some supplemental light from the early morning sky.

(left) A single fan of multi-colored aurora sweeps delicately across the intense star-filled night, lasting only a few seconds in Denali State Park.

(facing page) A rayed aurora dances above the "Million Dollar Bridge" across the Copper River northeast of Cordova as the moon casts its light on the river's surface.

About the author/photographers

A writer, photographer and broadcaster, **Cary Anderson** has covered news and features for many of the world's leading media organizations. His photo credits include *National Geographic, Newsweek, Popular Photography, Astronomy* and many other publications. He has written and narrated well over a thousand reports for the CBS Radio Network.

Anderson began to pursue his interests in nature photography in 1985, a year after his move from Washington State to Alaska. Although he has photographed wild subjects ranging from insects to polar bears, his work evolved to specialize in eagles and auroras.

Anderson's previous books include *Valley of the Eagles* (Fathom Publishing, 1995), *Alaska's Magnificent Eagles* (Alaska Geographic Society, 1997), and *The Eagle Lady* (Eagle Eye Pictures, 2003).

A collection of photographs and books by Cary Anderson may be viewed at his website: www.caryanderson.com

Dave Parkhurst is recognized by his peers as one of the world's leading natural phenomenon photographers, wholly dedicated to the pursuit of filming the aurora borealis since 1980. Parkhurst has spent countless thousands of hours working in sub-zero temperatures working to capture auroras on film. His photographs of the northern lights have appeared in numerous publications worldwide.

Every day during the summer, visitors to Anchorage can view his fascinating show, *Aurora: Alaska's Great Northern Lights*, in the Sydney Laurence Theater at the Alaska Center for the Performing Arts. In winter, he leads Aurora Watch® Tours for those wanting a personal aurora experience.

Born in New Orleans, Parkhurst grew up on the Islands of Sumatra and Java in Indonesia, and the Republic of Singapore before moving to Hawaii. Married with twin daughters, he has lived in Alaska since 1979. More of Dave Parkhurst's work can be seen at his website: www.thealaskacollection.com

70

Photographer Dave Parkhurst spending another cold winter night photographing auroras in Alaska.

Bibliography

Akasofu, Syun-Ichi, in collaboration with Jack Finch and Jan Curtis. *Secrets of the Aurora Borealis*. Penny Rennick, Editor. Anchorage: Alaska Geographic, 2002.

Brekke, Asgeir, and Alv Egeland. *The Northern Light: From Mythology to Space Research*. Berlin: Springer-Verlag, 1983.

Davis, Neil. *The Aurora Watcher's Handbook*. Fairbanks: University of Alaska Press, 1992.

Eather, Robert H. *Majestic Lights, The Aurora in Science, History and the Arts*. Washington, D.C.: American Geophysical Union, 1980.

Falck-Ytter, Harald. *Aurora: The Northern Lights in History, Mythology and Science*. Hudson, New York: Bell Pond Books, 1999.

Petrie, William. Keoeeit, *The Story of the Aurora Borealis*. Oxford, England: Pergamon Press, Ltd., 1963.

Savage, Candace. *Aurora: The Mysterious Northern Lights*. Vancouver, British Columbia: Greystone Books, 1994.

Sheet music published for the 1905 Lewis & Clark Centennial in Portland, Oregon included a song called "The Northern Lights Waltz."

Photo credits

Cary Anderson: pp. 4, 7, 9, 10, 11, 13, 15, 20, 21, 22, 23, 25, 28, 29, 37, 38, 39, 43, 55, 56, 60 (left), 66, 68, 70. Aurora collectibles from the author's collection.

Dave Parkhurst: pp. 1, 6, 14, 26, 33, 34, 42, 45, 49, 51, 54, 58, 59, 60 (right), 61, 62, 63, 64, 69, 72.

Pages 2, 3, and 31: Polar aurora animation by Colin Lamar, ©Dave Parkhurst.

Page 9: Ellen Anderson.

Page 12: Brian Anderson

Pages 12,17: Thanks to Debra Gust, researcher, Curt Teich Postcard Archives at the Lake County Discovery Museum, Wauconda, Illinois.

Page 30: Aurora height chart © Dave Parkhurst.

Page 40: (left) Courtesy Quidi Vidi Brewing Co. Ltd., (right) courtesy Aurora Vending.

Page 41: Coffee Safari label art ©Jon Van Zyle, used with permission.

Page 44: NASA photo by Robert Overmyer.

Page 57: Extremely Alaska Poster courtesy of Nicholas Jacobs, Fairbanks Conventon and Visitors Bureau.

Page 65: Iditarod pin ©1987 ITC, Rondy pin ©1995 Anchorage Fur Rendezvous Inc., both used with permission.

Page 67: Courtesy Space Telescope Science Institute (STScI). Still images of the aurora over Saturn taken by the Hubble Telescope: Saturn Aurora - HST.STIS. January 7th, 1998 - J. Trauger (JPL) and NASA. This image was created with support to the Space Telescope Science Institute, operated by the Association of Universities for Research in Astronomy, Inc., from NASA contract NAS 5-26555 and is reproduced with permission from AURA/STScI.

Ghostly pastel curtains of aurora pulsate above the Alaska Range in this view from Talkeetna. Autumn colors glow below in soft moonlight.